W9-AWR-218

PATTERNS OF LIFE

Gathering Food

Gathering Food

Daphne Butler

RAINTREE STECK-VAUGHN PUBLISHERS
The Steck-Vaughn Company
Austin, Texas

Published by Raintree Steck-Vaughn Publishers,
an imprint of Steck-Vaughn Company

Design: SPL Design

Library of Congress Cataloging-in-Publication Data

Butler, Daphne. 1945–
Gathering food / Daphne Butler
p. cm. -- (Patterns of Life)
Includes index.
Summary: Discusses the foods that different kinds of animals--and humans--eat and how they get them..
ISBN 0-8172-4200-7
1. Animals--Food--Juvenile literature. 2. Food--Juvenile literature. [1. Animals--Food habits. 2. Food.]
I. Title. II. Series: Butler, Daphne, 1945–
Patterns of life.
QL756.5.B87 1996
591.53--dc20

95-16626
CIP
AC

Printed and bound in Singapore
by KHL Printing Co Pte Ltd
1 2 3 4 5 6 7 8 9 0 99 98 97 96 95

Photographs: Zefa except for
NHPA (cover, title page, 8tr, 8bl, 13, 14tr, 17, 19, 21t, 21bl, 22, 25)
Robert Harding (8tl)

Contents

Going Shopping

In grocery stores and supermarkets you can buy many kinds of food. Which food is your favorite?

How many times a day do you need to eat? All creatures need to eat regularly to stay alive.

On the Farm

Some people do grow food in their gardens. Most food is grown on farms.

There are dozens of farm crops. Some, like wheat, are very important. At harvest time, farmers gather their crops. They often use special machines.

Animals are also kept on farms. What kind of foods do we get from animals?

From the Factory

Beside fresh food, we also eat food out of cans, foil packs, or bags from the freezer.

These are prepared foods. They have been cooked, preserved, or prepared in some way. This special work is done in a factory.

How much of your food is prepared before you buy it?

Gone Fishing

Sometimes we eat fish. Do you like fish? Have you ever been fishing?

People are not the only animals that go fishing. What other animals do you know that fish?

Hunters

Many animals prey on other animals for food. They attack quickly. These animals use sharp claws and teeth or beaks. Sometimes they hunt in a pack and share the meal.

Other animals may break open nests. They steal eggs or eat larvae.

What kind of meat do you eat? Where does it come from?

Waiting to Ambush

Some animals cannot move as quickly as their prey. These hide and wait, ready to attack. They stay still and pounce at the right time.

Some can even change the color of their skins as a disguise.

Feeding in Water

Some animals eat small creatures and plants in the water.

Each species, or kind of animal, has its own way of trapping its food. Birds use their beaks. What do other animals use?

Plant-Eaters

Many animals feed on plants. Their teeth are just the right shape. They nibble on green shoots, vegetables, or fruit.

Different species, or types of animals, eat different plants. They must live where the plants grow.

What kinds of plants do you eat? Where do they grow?

In Open Country

Some animals like to live together in herds. They have long legs and can run fast in open country. They feed on grass and the leaves of bushes and trees.

To get enough food, these animals must spend most of their time eating. What other animals do you know that eat grass?

Do you eat grass or leaves?

Working Together

Some creatures work together to collect food and store it for the winter. They live in a colony and share their supplies of food.

Both bees and ants work together this way.

What do people do? How do they share their food?

Animals That Clean Up

Some creatures live off rotting plant material and the remains or droppings of other animals.

Animals like worms, dung beetles, and vultures clean up the environment and take away the useless waste.

People make a lot of trash.

What happens to your leftover food? Who takes away your trash?

Useful Words

carnivores Animals that eat meat. Lions, owls, and sharks are carnivores.

harvest time The time when farmers gather their crops.

herbivores Animals that eat plants. Examples are donkeys and elephants.

omnivores Animals that eat both plants and meat. Dogs, many birds, and people are omnivores.

vegetarians People who choose to eat no meat. Many eat only fruit, vegetables, and plant products.

species A group of animals or plants that are the same kind or type. A tiger is one species of animal.

Index